Abdelhafid Mimouni

Maîtriser l'Hypertension: Comprendre, Prévenir, Agir

Abdelhafid Mimouni

Maîtriser l'Hypertension: Comprendre, Prévenir, Agir

ScienciaScripts

Imprint

Cover image: www.ingimage.com

This book is a translation from the original published under ISBN 978-620-6-71444-6.

Publisher:
Sciencia Scripts
is a trademark of
Dodo Books Indian Ocean Ltd. and OmniScriptum S.R.L publishing group

120 High Road, East Finchley, London, N2 9ED, United Kingdom
Str. Armeneasca 28/1, office 1, Chisinau MD-2012, Republic of Moldova, Europe
Printed at: see last page
ISBN: 978-620-8-04094-9

Controlling Hypertension: Understand, Prevent, Act

Author: Dr Abdelhafid Mimouni is an independent researcher specialising in the chemistry of bioinorganic systems, with extensive expertise in macromolecular synthesis and characterisation. He obtained his PhD in chemistry from the University of Paris XII in 1997 and a Diplôme des Etudes Approfondies des systèmes bioinorganiques from the University of Paris XI in 1993.

Summary: This book takes an in-depth look at hypertension (high blood pressure), a common and often silent medical condition. It begins with a general presentation of hypertension, highlighting its significant impact on cardiovascular health. The biological and physiological mechanisms of hypertension are detailed, highlighting the crucial role of the renin-angiotensin system and the vascular endothelium. The impact of diet is examined, with a focus on salt, saturated fats, fruit, vegetables and sugary drinks. Lifestyle factors, including physical activity, stress management and smoking cessation, are also discussed for their influence on blood pressure. In conclusion, this book offers recommendations for the prevention and management of hypertension, encouraging healthy lifestyle habits and regular medical check-ups to improve overall cardiovascular health.

Table of Contents

Introduction 3

Chapter 1: Understanding Arterial Hypertension 6

Chapter 2: Diet and Hypertension 11

Chapter 3: Lifestyles and Hypertension 15

Chapter 4: Traditional practices and natural remedies 19

Chapter 5: Prevalence of Hypertension 23

Chapter 6: Prevention and management of hypertension 26

Conclusion 35

References 39

Lexicon 42

Introduction

High blood pressure (hypertension) is one of the most common chronic diseases in the world, affecting millions of people of all ages and from all walks of life. It is characterised by excessively high blood pressure in the arteries, which can lead to a host of serious complications, including cardiovascular disease, stroke and kidney failure.

Understanding the mechanisms of hypertension is essential for several reasons. Firstly, a better understanding enables effective prevention strategies to be put in place, aimed at reducing the incidence of this disease. Secondly, proper management of hypertension can significantly improve the quality of life of those affected, by reducing the risk of serious complications. Finally, educating the public and healthcare professionals about the importance of managing hypertension helps to promote public health and reduce healthcare costs.

High blood pressure does not develop in isolation; it is often influenced by various environmental, dietary and genetic factors. Excessive salt consumption, diets high in saturated fats and processed foods, and lack of physical activity are all factors that can contribute to the development of this disease. In addition, socio-economic aspects, such as limited access to healthcare and nutritional information, may also play a crucial role in the prevalence of hypertension.

This book aims to provide a comprehensive overview of high blood pressure, exploring its causes, consequences and ways of preventing and managing it. Through a detailed analysis of current evidence, case studies and practical recommendations, we hope to provide valuable tools to help individuals better understand and control this insidious disease. By addressing the different facets of hypertension, we emphasise the importance of a healthy, balanced lifestyle, as well as the need for appropriate medical management to prevent long-term complications.

Chapter 1: Understanding Hypertension

Definition of hypertension

Hypertension is a medical condition characterised by chronically high blood pressure. Blood pressure is measured in millimetres of mercury (mmHg) and is made up of two values: systolic pressure (the pressure when the heart beats) and diastolic pressure (the pressure when the heart is at rest between beats). Blood pressure is considered high when the systolic pressure is above 140 mmHg and/or the diastolic pressure is persistently above 90 mmHg.

Biological and physiological mechanisms of hypertension

Hypertension results from a complex imbalance between several blood pressure regulation systems, involving hormonal, neurological and vascular factors. Here is an explanation of the main biological mechanisms involved in hypertension:

1. renin-angiotensin-aldosterone (RAA) system :

-Renin: Renin is an enzyme produced by the kidneys in response to reduced renal blood flow, low sodium concentration or stimulation of the sympathetic nervous system. Renin cleaves angiotensinogen (produced by the liver) into angiotensin I.

Angiotensin II: Angiotensin I is converted into angiotensin II by the angiotensin-converting enzyme (ACE), mainly in the lungs. Angiotensin II is a powerful vasoconstrictor that increases blood pressure by causing the smooth muscles of the blood vessels to contract. It also stimulates the secretion of aldosterone by the adrenal glands.

-Aldosterone: Aldosterone increases the reabsorption of sodium and water by the kidneys, thereby increasing blood volume and blood pressure.

2. cortisol :

-Cortisol, a hormone produced by the adrenal glands in response to stress, can also influence blood pressure. Cortisol increases the sensitivity of blood vessels to vasoconstrictors such as angiotensin II and noradrenaline. It can also increase blood volume by promoting sodium and water retention.

3 Sympathetic nervous system :

-Activation of the sympathetic nervous system leads to the release of noradrenaline, which causes the smooth muscles of the blood vessels to contract, thereby increasing peripheral vascular resistance and blood pressure. Chronic activation of this system can lead to persistent hypertension.

4 Endothelial dysfunction :

-The vascular endothelium plays a crucial role in regulating blood pressure by producing vasodilatory substances (such as nitric oxide) and vasoconstrictors (such as endothelin). Endothelial dysfunction, often due to risk factors such as smoking, obesity and hypercholesterolaemia, can lead to a preponderance of vasoconstriction and a rise in blood pressure.

General risk factors

Hypertension is influenced by a combination of genetic, environmental and dietary factors.

1 Genetic factors :

-Family history of hypertension

-Genetic variations influencing hormonal systems and salt sensitivity

2 Environmental factors :

-Chronic stress

-Exposure to noisy or polluted environments

3 Food factors :

-Excessive salt consumption

-Diets rich in saturated fats and sugars

-Low consumption of fruit, vegetables and fibre

-Excessive alcohol consumption

By understanding the biological and physiological mechanisms underlying hypertension and the associated risk factors, it is possible to develop effective prevention and management strategies. These include lifestyle modifications, a balanced diet, regular physical activity and, where necessary, drug treatment to maintain blood pressure at healthy levels and prevent serious complications.

Chapter 2: Diet and Hypertension

Impact of salt consumption on blood pressure

Salt, or sodium chloride, plays a major role in regulating blood pressure. Excessive salt consumption is strongly associated with an increase in blood pressure. The underlying mechanism involves an increase in blood volume: a high salt intake leads to increased retention of sodium and water by the kidneys, thereby increasing blood volume and, consequently, the pressure exerted on artery walls.

Epidemiological studies have shown that populations with a high salt intake have a higher prevalence of hypertension. Conversely, reducing salt intake can significantly lower blood pressure in both hypertensive and normotensive individuals. Current World Health Organisation (WHO) recommendations recommend limiting salt consumption to less than 5 grams a day to help prevent hypertension and associated cardiovascular disease.

The role of saturated fats and fried foods

Saturated fats and fried foods are dietary elements that can contribute to the development of high blood pressure. Saturated fats, found in large quantities in fatty meats, whole dairy products and processed foods, can increase blood cholesterol levels, thereby promoting atherosclerosis, a condition in which the arteries become rigid and narrow. This stiffening and narrowing of the arteries increases vascular resistance, leading to a rise in blood pressure.

Fried foods, often rich in saturated and trans fats, also contribute to obesity, a major risk factor for hypertension. In addition, cooking methods involving high temperatures can generate harmful compounds that have adverse effects on cardiovascular health.

The importance of fruit and vegetables in preventing hypertension

Fruit and vegetables play a crucial role in preventing and managing high blood pressure. They are rich in essential nutrients, particularly potassium, fibre and antioxidants, which help to regulate blood pressure. Potassium, in particular, helps to counteract the effects of sodium by relaxing the walls of the blood vessels and facilitating the excretion of sodium by the kidneys.

A diet rich in fruit and vegetables has been associated with a significant reduction in blood pressure in several clinical studies. Dietary recommendations encourage the consumption of at least five portions of fruit and vegetables a day to benefit from their protective effects against hypertension.

Effects of sugary drinks and alcohol

The consumption of sugary drinks is linked to an increase in blood pressure and the risk of obesity, two major risk factors for hypertension. Sugar-sweetened drinks, rich in added sugars, can lead to excessive weight gain, increasing the workload on the heart and raising blood pressure levels.

Alcohol, when consumed in excess, can also contribute to hypertension. Moderate alcohol consumption is defined as up to one drink a day for women and up to two drinks a day for men. Exceeding these amounts can lead to an increase in blood pressure. In addition, alcohol can interact negatively with antihypertensive drugs, reducing their effectiveness.

In conclusion, a balanced diet low in salt and saturated fats, rich in fruit and vegetables, and moderate consumption of sugary drinks and alcohol are essential for the prevention and management of hypertension. Adopting such dietary habits can not only help control blood pressure, but also improve overall cardiovascular health and reduce the risk of complications associated with hypertension.

Chapter 3: Lifestyles and high blood pressure

Influence of physical activity on blood pressure

Physical activity plays a crucial role in regulating blood pressure. Studies show that regular exercise can lower systolic and diastolic blood pressure by several points, thereby reducing the risk of developing cardiovascular disease. Aerobic exercise, such as brisk walking, running, swimming and cycling, is particularly effective. These activities strengthen the heart, improve blood circulation and encourage blood vessels to dilate, thereby reducing peripheral vascular resistance.

Regular exercise also helps maintain a healthy body weight, which is crucial for preventing and controlling hypertension. Obesity is a major risk factor for hypertension, and weight loss through exercise can lead to significant improvements in blood pressure.

Effects of stress and sedentary lifestyles

Chronic stress is a major contributor to high blood pressure. Stress triggers the release of cortisol and adrenaline, hormones that increase heart rate and constrict blood vessels, raising blood pressure. Stress management techniques such as meditation, yoga and deep breathing exercises can help reduce blood pressure by lowering cortisol levels and promoting relaxation.

Sedentary lifestyles also increase the risk of hypertension. Lack of physical activity can lead to weight gain, reduced insulin sensitivity and increased vascular resistance. Physical inactivity is often associated with an unhealthy diet and other risky behaviours, such as excessive alcohol consumption and smoking, exacerbating the risk of hypertension.

Smoking and hypertension

Smoking is a well-established risk factor for high blood pressure. The nicotine in cigarettes causes temporary vasoconstriction of the blood vessels, thereby increasing blood pressure. Smoking also damages artery walls, promoting atherosclerosis, which increases vascular resistance and blood pressure. In addition, smoking can reduce the effectiveness of antihypertensive drugs, further complicating the management of hypertension.

Smokers also have an increased risk of developing serious complications associated with high blood pressure, such as strokes, heart attacks and kidney disease. Quitting smoking is one of the most effective measures for reducing blood pressure and improving overall cardiovascular health.

Conclusion

Lifestyle habits play a crucial role in preventing and managing high blood pressure. Adopting a regular exercise routine, managing stress effectively, avoiding a sedentary lifestyle and quitting smoking are essential strategies for maintaining healthy blood pressure. These lifestyle changes, combined with a balanced diet and appropriate medical care, can significantly reduce the risk of complications associated with hypertension and improve the quality of life of those affected.

Chapter 4: Traditional practices and natural remedies

Traditional practices and the use of natural remedies have long been explored as alternative approaches to treating high blood pressure (hypertension). This chapter looks in detail at the effectiveness of these methods, with particular emphasis on garlic and other natural remedies.

Using Herbs and Traditional Remedies to Treat High Blood Pressure

In many cultures around the world, medicinal herbs have been used for their potential blood pressure lowering properties. For example, hawthorn (Crataegus spp.) is traditionally used in Europe for its beneficial effects on the cardiovascular system, including a slight reduction in blood pressure thanks to its flavonoids and procyanidins. St John's Wort (Hypericum perforatum) is known for its moderately hypotensive effects, probably linked to its active compounds such as hyperforin and hypericin.

Green tea (Camellia sinensis) contains catechins, powerful antioxidants that can help improve cardiovascular health by reducing oxidative stress and improving endothelial function. These herbs are often administered in the form of decoctions, infusions or concentrated extracts, and are incorporated into traditional diets to promote long-term cardiovascular health.

Effectiveness of Garlic and Other Natural Remedies

Garlic (Allium sativum) is one of the most widely studied natural remedies for its hypotensive effects. It contains a number of bioactive compounds, including allicin, which have been shown to have antihypertensive properties by relaxing blood vessels and reducing peripheral vascular resistance. Clinical trials have shown that regular consumption of garlic can result in a modest but significant reduction in blood pressure, making it a potential complement to conventional medical treatments.

Other natural remedies, such as ginkgo biloba, turmeric and omega-3 from fish oils, have also been shown to have beneficial effects on blood pressure regulation. Ginkgo biloba, for example, can improve peripheral blood circulation and vasodilation thanks to its flavonoids and terpenoids. Turmeric, with its active curcumin, has anti-inflammatory and antioxidant properties that could help reduce blood pressure by improving endothelial function.

Comparison with Modern Medical Treatments

Compared with modern medical treatments, such as angiotensin-converting enzyme (ACE) inhibitors, angiotensin II receptor blockers (ARBs) and diuretics, natural remedies often lack solid scientific evidence to support their long-term efficacy and safety. Modern medical treatments are based on rigorous clinical trials and meta-analyses that have established their effectiveness in reducing blood pressure and preventing cardiovascular complications.

Although promising, natural remedies are not generally recommended as a first-line treatment for patients with severe hypertension. They can be used as a complement to conventional medical treatments under the supervision of a healthcare professional, particularly for patients with mild to moderate hypertension or those seeking alternative options.

Conclusion

In conclusion, the approach of traditional practices and natural remedies to treat hypertension has interesting potential, but requires further studies to rigorously assess their efficacy and safety. An appropriate combination of these methods with conventional medical treatments could represent a beneficial integrative approach for certain patients, subject to appropriate medical advice and regular monitoring of blood pressure.

Chapter 5: Prevalence of Hypertension

This chapter explores the prevalence of high blood pressure (HBP) worldwide, analysing global statistics, trends and differences between developed and developing countries, as well as variations in different populations.

Global Statistics and Trends

Hypertension is one of the main causes of morbidity and mortality in the world. According to data from the World Health Organisation (WHO), approximately one adult in four worldwide suffers from hypertension. This prevalence varies considerably from one region of the world to another, and tends to increase with age.

Recent trends show an increase in the prevalence of hypertension, due in part to the rising prevalence of obesity, sedentary lifestyles and increased salt consumption in the diet.

Comparison between developed and developing countries

Developed countries generally have higher rates of hypertension than developing countries, due to factors such as rapid urbanisation, the adoption of Western lifestyles and the increasing prevalence of obesity. However, developing countries are also experiencing an alarming increase in the prevalence of hypertension due to nutritional and demographic transitions.

In developed countries, the prevalence of hypertension is estimated at around 30% in adults, while in some developing countries it can vary from 15% to 25%, but it is increasing rapidly.

Analysis of Rates of Hypertension in Various Populations

Rates of hypertension also vary within populations according to various factors, including age, sex, ethnic origin and socio-economic conditions. For example, populations of African and African-American origin have a higher prevalence of hypertension than populations of European origin.

Differences in dietary habits, levels of physical activity and access to healthcare also contribute to variations in hypertension rates between populations.

Conclusion

In conclusion, the prevalence of hypertension varies considerably worldwide, with increasing trends in many regions. Understanding these variations is essential to developing effective strategies for the prevention and management of hypertension worldwide. An integrated approach, taking account of socio-economic, cultural and environmental factors, is needed to combat this major cardiovascular disease.

Chapter 6: Prevention and management of hypertension

This chapter takes an in-depth look at prevention strategies, the importance of regular screening and medical check-ups, as well as modern treatment options and access to care for the management of high blood pressure (hypertension).

Strategies for Prevention through Diet and Lifestyle

Prevention of hypertension is based on adopting a healthy lifestyle, including a balanced diet and physical activity. The DASH (Dietary Approaches to Stop Hypertension) diet, rich in fruit, vegetables and whole grains, and low in salt, saturated fats and added sugars, is recommended for lowering blood pressure. This diet is particularly effective thanks to its high content of potassium, calcium, magnesium and fibre, all of which are linked to a reduction in blood pressure.

Reducing sodium consumption is a key component in the prevention of hypertension. Current recommendations suggest limiting sodium intake to less than 2.3 grams a day, equivalent to about one teaspoon of table salt.

Weight control is also essential, as obesity is a major risk factor for hypertension. Modest weight loss can have a significant effect on reducing blood pressure, often due to improved insulin sensitivity and reduced resistance to blood flow.

Importance of regular screening and medical check-ups

Regular Screenings

Regular blood pressure screening plays a crucial role in the early diagnosis of hypertension. Here are a few key points to underline their importance:

Early diagnosis: Regular screening can identify individuals with hypertension before significant cardiovascular damage occurs. Early diagnosis is essential to prevent serious complications such as stroke, heart attack and kidney failure.

Frequency of screening: Clinical guidelines recommend blood pressure measurements at least once a year for healthy adults. For those at increased risk of hypertension (for example, people with a family history, diabetic patients or those suffering from obesity), screenings should be carried out more frequently, often every three to six months.

Accessibility of Screenings: Screenings can be carried out in a variety of settings, including GP surgeries, pharmacies, and even at home using blood pressure monitoring devices. The availability of these options means that monitoring can be more regular and accessible to a greater number of people.

Education and awareness: Regular screenings provide an opportunity to educate patients about the risks associated with hypertension and the importance of maintaining normal blood pressure. This awareness can encourage patients to adopt healthier lifestyle behaviours.

Medical Controls

Regular medical check-ups are essential for the effective management of hypertension. Here are some important aspects of these checks:

Risk Factor Assessment: During medical check-ups, a detailed assessment of risk factors such as smoking, alcohol consumption, diet and physical activity is carried out. This assessment makes it possible to identify risky behaviours and provide personalised advice on how to modify them.

Comorbidity management: Many hypertensive patients also suffer from comorbidities such as diabetes, hypercholesterolaemia and obesity. Regular check-ups allow these conditions to be monitored and managed in a coordinated way, which is crucial for preventing cardiovascular complications.

Adaptation of treatment: Regular monitoring allows the treatment of hypertension to be adapted according to the patient's individual response. This includes adjusting drug dosages, switching to other classes of medication in the event of adverse effects, and implementing non-pharmacological strategies such as lifestyle modifications.

Minimising adverse effects: Regular medical check-ups enable potential adverse effects of antihypertensive treatments to be monitored. By identifying and treating these effects quickly, doctors can improve patient compliance and quality of life.

Ongoing Monitoring: Ongoing monitoring is essential to ensure that blood pressure remains under control over the long term. Regular appointments also provide an opportunity for patients to discuss concerns, ask questions, and receive ongoing support from their healthcare team.

In conclusion, regular screening and medical check-ups are essential elements in the management of high blood pressure. They enable not only early diagnosis and treatment, but also ongoing, personalised management of the condition, thereby improving long-term health outcomes.Modern Treatment Options and Access to Care

Treatment options for hypertension

Treatment options for arterial hypertension (AH) are vast and diverse, covering several classes of drugs that target different mechanisms of blood pressure regulation. Here is a more detailed description of the main classes of drugs used:

Diuretics:

Thiazides: These drugs, such as hydrochlorothiazide and chlorthalidone, work by increasing the excretion of sodium and water by the kidneys, which reduces blood volume and lowers blood pressure.

Loop diuretics: Drugs such as furosemide, which are more powerful than thiazides and are often used in cases of heart failure or severe oedema.

Potassium-sparing diuretics: Spironolactone and eplerenone, which help prevent excessive potassium loss, often used in combination with other diuretics.

Beta-blockers: Drugs such as atenolol and metoprolol, which reduce the heart rate and the force of contraction of the heart, thereby lowering blood pressure.

Angiotensin-converting enzyme inhibitors (ACEI): Drugs such as lisinopril and enalapril, which prevent the formation of angiotensin II, a chemical that constricts blood vessels. By reducing angiotensin II levels, these drugs allow blood vessels to relax and widen.

Angiotensin II receptor blockers (ARBs): Losartan and valsartan belong to this class of drugs, which directly block the action of angiotensin II on its blood vessel receptors, resulting in a decrease in peripheral vascular resistance.

Calcium channel blockers: Drugs such as amlodipine and diltiazem, which prevent calcium from entering the cells of the heart and blood vessels, leading to relaxation of the vessels and a reduction in blood pressure.

Direct renin inhibitors: Aliskiren, which directly inhibits the activity of renin, a key enzyme in the renin-angiotensin-aldosterone system, thereby helping to regulate blood pressure.

Access to healthcare

Access to healthcare remains a major challenge, particularly in low-income regions and among disadvantaged populations. Several strategies can be implemented to improve access to treatment for hypertension:

Strengthening community health systems: By developing basic health infrastructure and supporting community health centres, primary healthcare can be made more accessible and affordable.

Training healthcare professionals: It is crucial to train healthcare professionals, including doctors, nurses and community health workers, in the effective management of hypertension. This includes recognising symptoms, administering appropriate treatments and raising awareness of preventive measures.

Use of telehealth technologies: Telehealth technologies allow patients to be monitored remotely, reducing the need to travel and enabling continuous monitoring and rapid intervention. Virtual consultations and mobile applications for blood pressure management are examples of tools that can improve access to care.

Awareness and education programmes: Educating the public about the risks of hypertension, the importance of regular blood pressure monitoring, and the lifestyle changes needed can help prevent and manage the condition. Awareness campaigns can be conducted through local media, community workshops and partnerships with local organisations.

Public health policies: Governments and public health bodies can play a key role in implementing policies to improve access to treatment for hypertension. This can include funding health programmes, reducing the cost of antihypertensive drugs, and promoting healthy lifestyles through national initiatives.

By combining these approaches, it is possible to make significant progress in the management of hypertension and improve the quality of life of people affected by this condition.

Conclusion

In conclusion, the prevention and management of hypertension requires a multi-dimensional approach, combining healthy lifestyle strategies, regular screening and effective management by healthcare professionals. Progress in understanding the pathophysiological mechanisms of hypertension has led to the development of effective treatments that can significantly reduce the risk of associated cardiovascular complications. An integrated and personalised approach is essential to improve long-term outcomes for patients with hypertension.

Conclusion

High blood pressure (hypertension) is a common chronic medical condition affecting millions of people worldwide. This book has explored in depth the different aspects of hypertension, from its definition to its biological mechanisms, from risk factors to the impact of diet and lifestyle on this major cardiovascular disease.

Summary of Chapters

In the introduction, we highlighted the importance of understanding and managing hypertension, because of its significant implications for cardiovascular health and quality of life. We examined how hypertension is a silent condition that can go undetected for years, increasing the risk of serious cardiovascular diseases such as stroke and heart attack.

Chapter 1 laid the foundations by defining hypertension and exploring the biological and physiological mechanisms underlying this disease. We discussed the crucial roles of the renin-angiotensin system, the vascular endothelium, and the mechanisms of blood pressure regulation.

Chapter 2 focused on the impact of diet on hypertension, highlighting the effect of salt, saturated fat, fruit and vegetables, as well as sugary drinks and alcohol on blood pressure. We examined how a diet rich in fruit and vegetables, such as the DASH diet, can not only prevent but also treat hypertension.

In Chapter 3, we explored the importance of lifestyle factors, including physical activity, stress management, smoking cessation and sedentary lifestyle, in the prevention and management of hypertension. We have highlighted how healthy lifestyle habits can help to lower blood pressure and improve overall cardiovascular health.

Future prospects and recommendations

As we conclude this book, it is essential to recognise that hypertension is a complex disease requiring a multi-dimensional approach. Clinical guidelines and recommendations for the management of hypertension, such as those issued by the ESC/ESH, provide an essential framework for clinicians and patients in the prevention, diagnosis and treatment of this disease.

Advances in medical research continue to provide new insights into the pathophysiology of hypertension and potential therapeutic approaches. Innovation in pharmacological treatments, medical devices and lifestyle interventions offers new promise for improving clinical outcomes and reducing the prevalence of hypertension worldwide.

Call to Action

Finally, this book aims to raise awareness and educate people about the importance of preventing and managing hypertension. It is crucial for everyone to be aware of their own risks, adopt healthy lifestyle habits and seek regular medical follow-up to prevent the potentially serious complications of hypertension.

Together, we can make progress in the fight against hypertension and improve cardiovascular health worldwide.

References :

1. Kearney, P. M., Whelton, M., Reynolds, K., Muntner, P., Whelton, P. K., & He, J. (2005). Global burden of hypertension: Analysis of worldwide data. The Lancet, 365(9455), 217-223. doi:10.1016/S0140-6736(05)17741-1
2. Williams, B., Mancia, G., Spiering, W., Rosei, E. A., Azizi, M., Burnier, M., ... Zanchetti, A. (2018). 2018 ESC/ESH Guidelines for the management of arterial hypertension. European Heart Journal, 39(33), 3021-3104.
3. Chobanian, A. V. (2007). Clinical practice: Isolated systolic hypertension in the elderly. New England Journal of Medicine, 357(8), 789-796. doi:10.1056/NEJMcp071137
4. Harrison, D. G., & Gongora, M. C. (2009). Oxidative stress and hypertension. Medical Clinics of North America, 93(3), 621-635.
5. L. J., Moore, T. J., Obarzanek, E., Vollmer, W. M., Svetkey, L. P., Sacks, F. M., ... Bray, G. A. (1997). A clinical trial of the effects of dietary patterns on blood pressure. New England Journal of Medicine, 336(16), 1117-1124.
6. He, F. J., & MacGregor, G. A. (2009). A comprehensive review on salt and health and current experience of worldwide salt reduction programmes. Journal of Human Hypertension, 23(6), 363-384.
7. Pescatello, L. S., Franklin, B. A., Fagard, R., Farquhar, W. B., Kelley, G. A., & Ray, C. A. (2004). American College of Sports

Medicine position stand: Exercise and hypertension. Medicine & Science in Sports & Exercise, 36(3), 533-553.

8. Steptoe, A., Kivimäki, M., & Stress, Health, & Aging Research Programme. (2012). Stress and cardiovascular disease: An update on current knowledge. Annual Review of Public Health, 33, 11-26.
9. Chobanian, A. V. (2007). Clinical practice: Isolated systolic hypertension in the elderly. New England Journal of Medicine, 357(8), 789-796.
10. Williams, B., Mancia, G., Spiering, W., Rosei, E. A., Azizi, M., Burnier, M., ... Zanchetti, A. (2018). 2018 ESC/ESH Guidelines for the management of arterial hypertension. European Heart Journal, 39(33), 3021-3104.
11. Touyz, R. M., & Montezano, A. C. (2018). Angiotensin II and vascular injury. Current Hypertension Reports, 20(1), Article 39.
12. Harrison, D. G., & Gongora, M. C. (2009). Oxidative stress and hypertension. Medical Clinics of North America, 93(3), 621-635.
13. Daviglus, M. L., Pirzada, A., & Talavera, G. A. (2014). Cardiovascular risk factor awareness and management in ethnic minorities in the United States. Progress in Cardiovascular Diseases, 57(3), 240-250.
14. NCD Risk Factor Collaboration (NCD-RisC). (2016). Worldwide trends in blood pressure from 1975 to 2015: A pooled analysis of

1479 population-based measurement studies with 19-1 million participants. The Lancet, 389(10064), 37-55.

15. Brown, M. J., & McInnes, G. T. (2017). Management of hypertension: An update. Clinical Medicine, 17(1), 13-18.

16. Ried, K., Frank, O. R., & Stocks, N. P. (2009). Aged garlic extract lowers blood pressure in patients with treated but uncontrolled hypertension: A randomised controlled trial. Maturitas, 67(2), 144-150.

17. Xin, X., Wei, J., Shen, L., & Lei, Y. (2017). Efficacy of traditional Chinese medicine in the treatment of essential hypertension: A meta-analysis. Evidence-Based Complementary and Alternative Medicine, 2017, Article ID 5157469. doi:

18. Mills, K. T., Bundy, J. D., Kelly, T. N., Reed, J. E., Kearney, P. M., Reynolds, K., ... He, J. (2016). Global disparities of hypertension prevalence and control: A systematic analysis of population-based studies from 90 countries. Circulation, 134(6), 441-450.

19. Zhou, D., Xi, B., Zhao, M., Wang, L., & Veeranki, S. P. (2016). Uncontrolled hypertension increases risk of all-cause and cardiovascular disease mortality in US adults: The NHANES III Linked Mortality Study. Scientific Reports, 6, Article 39394.

20. Whelton, P. K., Carey, R. M., Aronow, W. S., Casey, D. E., Collins, K. J., Dennison Himmelfarb, C., ... Wright, J. T. (2018). 2017

Lexicon

-Hawthorn (Crataegus spp.): Medicinal plant traditionally used for its beneficial effects on the cardiovascular system.

-Diuretic: Drug which increases the excretion of urine and sodium by the kidneys, thereby reducing blood pressure.

-DASH (Dietary Approaches to Stop Hypertension): A dietary approach aimed at reducing blood pressure by promoting a diet rich in fruit, vegetables and whole grains, and low in salt, saturated fats and added sugars.

-DASH diet: Diet based on the recommendations of the DASH approach, known for its beneficial effects on reducing blood pressure.

-Renin-angiotensin-aldosterone system: Hormonal system which regulates blood pressure by modifying blood volume and vascular resistance.

-Aldosterone: Steroid hormone produced by the adrenal glands, playing a role in regulating blood pressure by increasing sodium reabsorption in the kidneys.

-Angiotensin: Vasoconstrictor peptide hormone which regulates blood pressure by increasing peripheral vascular resistance.

-Endothelium: Layer of cells lining the blood vessels, playing a key role in regulating vascular tone.

-Arterial hypertension: abnormal and persistent increase in blood pressure in the arteries.

-Renin-angiotensin system: Hormonal system involved in regulating blood pressure via angiotensin II and aldosterone.

-Endothelium: Layer of cells lining the blood vessels, playing a key role in regulating vascular tone.

-Physical activity: Regular physical exercise helps to reduce blood pressure by various mechanisms, including reducing vascular resistance.

Printed by Books on Demand GmbH, Norderstedt / Germany